IF DINOSAURS WERE HERE TODAY
SNEAKY HUNTERS

Copyright © 2024 Bright Bound Ltd

First published in 2024 by Hungry Tomato Ltd
F15, Old Bakery Studios, Blewetts Wharf, Malpas Road, Truro, Cornwall,
TR1 1QH, UK.

No part of this publication may be reproduced, stored in a retrieval system, or transmitted in any form or by any means, electronic, mechanical, photocopying, recording, or otherwise, without prior written permission of the copyright owner.

A CIP catalogue record for this book is available from the British Library.

ISBN 9781916598928

Printed in China

Discover more at
www.hungrytomato.com

Picture Credits:
(abbreviations: t=top, b=bottom, m=middle, l=left, r=right, bg=background)

Alamy: World foto 7tl. American Museum of Natural History: Francois Gohier 7mr. Andrew kerr 26bl. Corbis: Ian Hodgson/Reuters 9ml, 18-19bg, 24bl. Fossil Finds: Stone 7bl. Leonello Calvetti 22tl, 25tr. Nature PL: Anup Shah 1bg, 8br, 14-15bg; Jurgen Freund 9br, 16-17bg. NHPA: Stephen Dalton 20-21bg, 28bl; John Shaw 4bg, 9bl, 10-12bg. Shutterstock: 3l, 5b, 9tr, 12-13bg, 23tr, 26tr; Catmando 8ml; David.costa.art 6tl; Dennis—S 6mr, 9mr; Dotted yeti 28tl; Erhan Inga 25br; Gorodenkoff 30tr; Herschel Hoffmeye 27tl; iurii 29bl; Jaroslav Moravick 31tl; Kamomeen 23ml; Kostiantyn Ivanyshen 8tl, 8mr; LuFeeTheBear 29mr; MaHiATH 22b; Maksim Shchu 6bl, 29tl, 30bl; Metha1819 23br; Pecold 31mr; Penny Hicks 31br; Sanut Fuanignakhon 16br; SHELIAKIN MAKSIM 25ml; rodos studio FERHAT CINAR 27mr.. Simon Mendez 1bg, 7tl, 7mr, 8br, 9ml, 14-15bg, 18-19bg, 24bl.

Every effort has been made to trace the copyright holders, and we apologise in advance for any unintentional omissions. We would be pleased to insert the appropriate acknowledgements in any subsequent edition of this publication.

IF DINOSAURS WERE HERE TODAY!
SNEAKY HUNTERS

by John Allan
Illustrated by Simon Mendez

HUNGRY TOMATO

WARNING! These extinct beasts are not alive today. But just imagine if they were...

CONTENTS

The Story of the Dinosaurs	6
Timeline	8
Big Claws, Narrow Jaw - *Baryonyx*	10
Small but Smart - *Troodon*	12
Predator or Prey? - *Ceratosaurus*	14
Bird or Dinosaur? - *Oviraptor*	16
Super Sprinters - *Struthiomimus*	18
Scary Scavengers - *Coelophysis*	20
Hunting Strategies	22
The Lives of Dinosaurs Today	24
Did You Know?	26
True or False?	28
Uncovering the Past	30
Index & Glossary	32

Words in **BOLD** can be found in the glossary.

THE STORY OF THE DINOSAURS

Planet Earth is around 4.5 billion years old. Rocks containing traces of living things shows us that there's been life on Earth for around 3.6 billion years. During Earth's long history, the planet and the creatures that roam it have changed drastically. We've all heard of the dinosaurs, but where did they come from, and where are they now?

WHEN DINOSAURS ROAMED

Dinosaurs were the most famous and fascinating animals to come from these prehistoric times. Dinosaurs were the biggest land-living creatures to have ever lived. Alongside these giants lived smaller, bird-like dinosaurs, flying reptiles and huge ocean-dwelling beasts.

Then, 65 million years ago, the dinosaurs were suddenly gone! Scientists believe that a huge asteroid hit Earth, wiping out most living things. The extinction of the dinosaurs allowed for the rise of new animals: 4 million years ago, humans appeared!

FOSSIL FINDS

Humans began observing rocks and, in the 1700s, discovered that the **fossils** they contained were the remains of ancient plants and animals. Fossil hunting became popular and the study of fossils – **palaeontology** – was born.

In 1842, scientist Sir Richard Owen invented the term 'Dinosauria' to describe the giant creatures that had once walked the Earth. Their remains fascinated both scientists and ordinary people – everyone wanted to know what these creatures had been like.

THEN AND NOW

For over two centuries, dinosaurs have amazed and fascinated us. We wonder how they'd compare to the animals of today.

Could prehistoric meat-eaters adapt their hunting style to the prey that's available today?

Could the Struthiomimus outrun today's fastest predator – the cheetah?

And would some dinosaurs like Troodon and Coelophysis be classed as vermin, like rats?

THE UNKNOWN

We may never know exactly what it would be like to live with dinosaurs. We can only imagine, taking what we've discovered from their fossilised remains, and comparing it to what we know about modern animals to picture what life would be like if dinosaurs were here today!

If you've got the courage, read on...
...be prepared for some truly bizarre and spine-tingling - though imaginary - encounters between human or animal and beast.

TIMELINE

TRIASSIC PERIOD
[252-201 MILLION YEARS AGO]

Dinosaurs appeared towards the end of the Triassic **period**. They tended to live by the seaside, along riverbanks and in desert **oases** where water was plentiful. Early dinosaurs included Plateosaurus and Coelophysis.

CRETACEOUS PERIOD
[145-66 MILLION YEARS AGO]

This was when some of the most famous dinosaurs lived, including T.rex, Triceratops and Spinosaurus. Who knows what other dinosaurs would have lived since then if they hadn't all been wiped out by the huge meteorite?

JURASSIC PERIOD
[201-145 MILLION YEARS AGO]

During the Jurassic period, Earth's climate became moister and milder – new plants and forests grew, meaning new food sources for plant-eating dinos. As a result, both plant- and meat-eating dinosaurs started to grow much bigger.

COELOPHYSIS

Name meaning 'hollow form'

These early, primitive dinos were small and lightly-built, with slim legs and a long neck and tail. This made them fast, agile hunters. They are thought to have lived worldwide in the early part of the Age of Dinosaurs.

CERATOSAURUS

Name meaning 'horned lizard'

This meat-eater had a distinctive look – with a horn on its snout, a pair behind its eyes, and a jagged crest down its back.

252 MILLION YEARS AGO

201 MILLION YEARS AGO

TROODON

Name meaning 'wounding tooth'

Unlike other meat-eaters, Troodons had sharp edges on their teeth, which is usually characteristic of herbivorous dinosaurs, suggesting they ate both meat AND plants!

OVIRAPTOR

Name meaning 'egg thief'

This feathery, turkey-sized dino has been discovered in Mongolia and Asia. Being from the Cretaceous period, it was likely one of the last dino species to develop.

MASS EXTINCTION

For millions of years, dinosaurs ruled the Earth, until there was a **mass extinction**. There is evidence that a **meteorite** struck Earth around 65 million years ago, exploding rock fragments, causing **tsunamis** and forest fires, resulting in the death of the dinosaurs and all other reptiles of the time.

BARYONYX

Name meaning 'heavy claw'

Named after its enormous thumb claws, Baryonyx was a fierce predator. Their fossils have been found in Southern England.

STRUTHIOMIMUS

Name meaning 'ostrich mimic'

The fossils of this small, two-footed dinosaur with a toothless jaw have been found in North America.

145 MILLION YEARS AGO

65 MILLION YEARS AGO

BIG CLAWS, NARROW JAWS
BARYONYX

A rush of water through the rapids and over the falls. The silvery leap of a salmon as it migrates. A snap! And it's caught in mid-air by an alert grizzly bear. Then another snap! And a Baryonyx has closed its jaws on the same fish. It's an unequal struggle, and the stronger dinosaur takes its prize away from the disappointed mammal.

When its skeleton was found, Baryonyx immediately attracted a huge amount of interest. Nothing like it was known. It had a big claw on its hand, and long narrow jaws with small crocodile-like teeth. When the fossils of fish remains were found in its stomach, scientists realised that it must have hunted fish like a grizzly bear, hooking them out of the water with its big claw or snapping them up in its big mouth. Baryonyx would take its place alongside bears, herons, and other fish hunters in modern streams and rivers.

BARYONYX
PRONOUNCED
ba-ree-on-icks

LIVED
Early Cretaceous period
130-125 million years ago

LENGTH
up to 10 metres (33ft)

DIET
Piscivore

FOSSIL FINDS

The first part of a Baryonyx skeleton to be found was a huge claw from the first finger of its hand. The claw was massive; its outer curve was over 30cm (12in) long! It would have been used to spear, catch and kill their prey. The rest of the skeleton was eventually found. In its stomach were fish bones and scales, showing that fish were its main food. There were also dinosaur bones in there, proving that Baryonyx ate bigger prey too.

LARGE FISH HUNTERS

None of the freshwater fish-hunting animals from our modern world – not even the largest grizzly bear – can begin to compare in size to Baryonyx.

SMALL BUT SMART
TROODON

Bounding across the outback comes a red kangaroo – its strong hind legs and balancing tail make it one of the fastest plant-eaters around. However, it's no match for the fast Troodon pair that converge, hounding it with their sharply-clawed hands, killer-clawed toes, and jagged, razor-sharp teeth. Even smaller than average, or young, Troodons would make tough predators for modern mammals thanks to their speed and fierceness.

By the Cretaceous period there were many fast-running plant-eaters. As a result, predators learned to run faster, in order to catch their speedy prey. Troodon's natural prey would have been the sprinting hypsilophodonts – fast plant-eaters that were about Troodon's size. In today's world, it would have been small plant-eating dinosaurs, mammals and lizards. Many scientists now think that Troodon also ate plants, making it an **omnivore**.

TROODON
PRONOUNCED
true-don

LIVED
Late Cretaceous period
77-65 million years ago.

LENGTH
2 metres (6.5ft)

DIET
Omnivore

BIG BRAINS
Troodon had the largest brain in relation to the size of its body out of all the dinosaurs. They were probably one of the smartest dinos of their time!

EGGS AND NESTS

Troodon is a close relative of birds. It made simple mud nests to lay its eggs, which were laid in pairs. This suggests that female Troodons had a pair of egg tubes in their bodies; whereas modern birds have just one. Female Troodons probably sat on their eggs to keep them warm and protect them from predatory dinos.

PREDATOR OR PREY?
CERATOSAURUS

With a row of sharp and bony horns on its head and along its back, Ceratosaurus was built to defend itself, but the big hunter is not always invulnerable. Even a creature as fierce as Ceratosaurus could find itself the victim of more powerful or better organised predators. An elderly or ill Ceratosaurus would be easy prey to a pride of hungry lions.

On the Jurassic plains, Ceratosaurus was a successful large hunter – a force to contend with among the long-necked **sauropods** and the two-footed ornithopods of the time. However, there were also other hunters around, such as Allosaurus which was bigger and stronger, and Ornitholestes which was smaller and probably hunted in packs. They would probably all be in competition for the same food. Ceratosaurus would probably be a protected animal today, despite being distrusted by people who live in its hunting area.

CERATOSAURUS

PRONOUNCED
seh-rat-uh-saw-rus

LIVED
Late Jurassic period
155-150 million years ago.

LENGTH
6 metres (20ft)

DIET
Carnivore

DINNER TIME
Ceratosaurus is thought to have hunted and scavenged a wide variety of animals, from plant-eating dinos to fish and reptiles. In fact, it may have had one of the most varied diets of any dinosaur.

BIG TEETH

The skull of Ceratosaurus is a lightweight structure made up of thin struts of bone surrounding huge holes – just like most dinosaur skulls. The feature that makes it different, apart from the horns, is the enormous size of its teeth which are far bigger in proportion to the size of the skull than those of other **theropods**.

BIRD OR DINOSAUR?
OVIRAPTOR

A quiet November morning on Christmas Island is interrupted by shrill cries from a flock of Oviraptors. The island's **juvenile** red crabs have left the safety of their forest home and begun their annual migration to the sea. For these feathered dinosaurs, Christmas has come early and they excitedly communicate news of the feast to each other.

Modern birds are very similar to the Oviraptor. However, they have some big differences. The dino's diet was much more varied because of its strong beak and powerful curved jaws. It could easily crunch through shell and bone. Shellfish like these crabs would be easy prey for the Oviraptor.

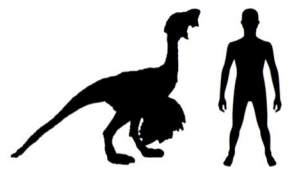

OVIRAPTOR
PRONOUNCED
ovi-rap-tor

LIVED
Late Cretaceous period
85-70 million years ago.

LENGTH
up to 2 metres (6.5ft)

DIET
Omnivore

EGG STEALER
Oviraptor was always thought to have been an egg-stealer, slinking around other dinosaur's nests and stealing their eggs and young. This was because the first Oviraptor skeleton found was close to a nesting site thought to belong to a different dino. 50 years later, an Oviraptor skeleton was found sitting on a nest of the same eggs. It had died in a sandstorm while incubating them. They were Oviraptor nests all along!

IS IT A BIRD?
Several of Oviraptor's physical features, such as its head crest and beak, have led scientists to believe that its closely related to modern birds, such as the Cassowary.

Cassowary

SUPER SPRINTERS
STRUTHIOMIMUS

The crowd cheers as the jockeys push the racehorses on and on. However, years of fitness training aren't enough to match the natural speed of the flock of Struthiomimus that have joined the race. These slender, long-legged dinosaurs effortlessly exceed speeds of 40 mph (65km per hour) – an essential survival skill for these animals that lived alongside giant carnivores such as Tyrannosaurus and Giganotosaurus.

The speed of Struthiomimus at a sprint may well have led to its inclusion in modern sporting fixtures: they were one of the fastest dinosaurs and could probably outrun many of today's animals. They lived and hunted in packs, and had large brains in relation to their body size, suggesting they weren't only fast runners, but clever too!

STRUTHIOMIMUS
PRONOUNCED
stroo-theo-me-mus

LIVED
Late Cretaceous period
76-65 million years ago.

LENGTH
2-4 metres (8-13ft)

DIET
Omnivore

BIRD LIKE
Struthiomimus resembled today's large flightless birds, such as ostriches and emus, although, unlike these creatures, they also had very long tails!

LONG TAIL
The dinosaur's long tail can clearly be seen in the image of a fossilised Struthiomimus (left). The long tail balanced the dinosaur as it ran at high speeds, and gave it the manoeuvrability to swerve, and turn quickly to avoid danger.

SCARY SCAVENGERS
COELOPHYSIS

In the darkness of a damp and smelly night, scavenging beasts sniff around the waste of a city. Foxes, once rural hunters, have become urban **scavengers**, finding food wherever it's discarded by people. But they're not alone – small dinosaurs such as Coelophysis, once rural hunters themselves, have also turned to scavenging food. These young dinosaurs are prepared to challenge the larger fox for the prize of a stolen meal.

Although Coelophysis had the build of a fast and agile hunter, it would've been perfectly happy to eat any dead food that happened to be lying around. Today it would be an urban scavenger and would therefore be in competition with the foxes and rats of our cities.

COELOPHYSIS
PRONOUNCED
see-low-fie-sis

LIVED
Late Triassic period
225-190 million years ago

LENGTH
up to 2.5 metres (8ft)

DIET
Carnivore

COMPETITIVE HUNTING
Coelophysis wasn't the only efficient meat-eating dinosaur around. Lots of the other meat-eaters were active, lightweight animals as well. They would all have competed for the same food, whether they were hunting for small living things like lizards or crocodiles, or scavenging from the corpses of dead animals.

SKILLFUL PACK HUNTERS
Coelophysis most likely hunted in packs like modern wolves and dogs. It wouldn't have eaten as much though, as mammals need more food to support their warm-blooded lifestyle.

HUNTING STRATEGIES

Whilst all meat-eating dinos had to hunt and catch their own food, they did this in different ways, using different strategies. The question is, could they hunt in the same way if they lived on Earth today? Would they just be hunting and eating their own kind or would they chase modern day animals and, of course, humans?

DIFFICULT DIETS

There's no doubt that most dinosaurs' diets and hunting strategies would be different if they existed today. However, the biggest meat-eaters, such as Tyrannosaurus, would probably have continued to hunt their traditional prey, the sauropods, as our smaller modern animals would be too small to fill them up (unless they ate A LOT).

They would also need to be able to run much faster to stand a chance of catching modern animals like zebras which are very quick and agile. Who knows if they would have been able to do this?

POWERFUL PACKS

Many dinosaurs lived and hunted in packs, which gave them safety in numbers and made them fierce predators. If farmers in Africa find wild dogs and lions a threat to their herds and flocks, then a prowling pack of dinos would be an even bigger problem!

LONE RANGERS

Prehistoric carnivores that hunted alone, like Ceratosaurus, would likely be tempted by a modern herd of domesticated animals, such as sheep, goats or cows, too. These dinos would be the perfect size for tackling a flock of farm animals alone. Herders would have to be armed with more than just sticks to keep these predators away.

HUNTING TIME!

Scientists think it's likely that dinosaurs hunted during the night as well as the day. They've found evidence that some dinos, including Troodon, had large eyes which suggests that they had very good eyesight, like today's owls and crocodiles. They would have used this to their advantage to sneak up on unsuspecting prey in the dark. It would certainly make you think better of going for a night-time dog walk if they were around today!

THE LIVES OF DINOSAURS TODAY

Both humans and dinosaurs would lead completely different lives if we existed alongside each other. Imagine walking through a field and seeing a massive flock of dinos being herded by a farmer, or finding a smaller dino, like Coelophysis, digging through your food scraps! What do you think would be the biggest change to your day-to-day life?

HUMAN IMPACT

Our farming and cities could have destroyed the natural environment of the big plant-eating dinosaurs to such an extent that they would be dying out. As a result, meat-eating dinosaur numbers would be declining, probably to such an extent that we would begin to feel their loss, and conservation groups would be lobbying to protect them, like they do for endangered species today.

CULTURE AND SOCIETY

We would probably have absorbed dinosaurs – even some of the meat-eaters – into our culture and civilisation, such as using Struthiomimus as a racing animal. What do you think would be more interesting to watch: horse racing or Struthiomimus racing?

VERMIN AND PESTS

Small dinosaurs might infest our cities, scavenging for food, breaking into buildings, and scaring cats and dogs. In fact, dinos like Troodon and Coelophysis would probably be classed as vermin, like rats, and measures would be taken to destroy them. If our civilisation had developed alongside them, we would think them very ordinary and a complete irritation.

DINOSAUR SAFARI

Would it be practical to keep such big animals in a zoo or a safari park? In the real world, we love to see big animals up close. Instead of a bear, imagine seeing a Baryonyx behind the bars. There would have to be lots of safety measures in place! If we kept dinosaurs in captivity, their specialist diet would be very important. Especially for the biggest sauropods – we would need to plant huge forests to give those humungous **herbivores** enough food to eat!

ROAST TROODON

Some of the smaller dinosaurs would have been used as our main food source, like chickens. Troodon would have made a great meat choice for a roast dinner! What do you think it would taste like?

DID YOU KNOW?

Dinosaurs are fascinating creatures. Scientists are constantly discovering more about them and finding answers to the world's most curious questions. Did you know these amazing facts about dinosaurs?

HOW FAR COULD A DINOSAUR SEE?

It depends. Hunting dinosaurs would have needed sharp eyesight to keep their prey in view. In some, their eyes were pointed forward, like ours, to help them to judge the distance to their prey. Plant-eaters would have been better off with eyes that could see around as far as possible, to detect danger coming from every direction.

DID DINOSAURS HAVE SLIT EYES?

Perhaps. If the active meat-eaters hunted like modern cats, they would probably have had slit eyes like cats, too. This would help them see in the dark, without being blinded by sunlight in the daytime.

WHAT WAS THE MOST BLOODTHIRSTY DINOSAUR?

Small meat-eaters would have to be very fierce if they were to kill something much bigger than themselves, so Velociraptor would probably qualify.

WERE ANY DINOSAURS CANNABALISTIC?

Possibly, although it's likely to have been very rare. For many years, scientists thought that Coelophysis ate its own kind, as bones were found in its stomach. However, these bones were examined further and found to be from prehistoric crocodiles.

HOW FAR COULD A DINOSAUR TRAVEL WITHOUT RESTING?

Many of the big plant-eating dinosaurs seem to have migrated hundreds of miles with the seasons.

TRUE OR FALSE?

There are a lot of fun facts about dinosaurs but what is actually true and what is a myth? Put your dino knowledge to the test with this true or false quiz.

THE BIGGEST DINOSAURS WERE MEAT-EATERS!

FALSE. The largest dinosaurs to exist belonged to a group called sauropods, which were plant-eaters.

DINOSAURS COULD SURVIVE IN MODERN DAY ENVIRONMENTS!

TRUE. If they had enough of the right things to eat, they probably would. Meat-eaters would be quite successful. Plant-eaters would probably find it more difficult because they were more selective about what they ate.

DINOSAURS ONLY HAD A FEW TEETH!

FALSE. Duck-billed dinosaurs such as Edmontosaurus had hundreds of teeth. However, a lot of these weren't used at the same time. Most grew upwards in stacks to replace those that were being worn away through use.

MOST DINOSAURS HAD THE SAME NUMBER OF BONES!

TRUE. All dinosaurs had pretty much the same number of bones (just as cats, dogs and horses do), but they differ a little by the number of **vertebrae** in the neck and tail.

WE WILL NEVER FIND ALL OF THE DINOSAUR FOSSILS!

TRUE. We're sure that only a fraction of those that ever existed have been discovered. The majority of them are buried deep in the earth. We will only ever be able to find the ones that lie closer to the surface.

WE COULD HAVE LIVED ALONGSIDE DINOSAURS!

FALSE. Mammals are only here because there are no dinosaurs. In the Age of Dinosaurs, mammals were small, like mice or opossum. There wasn't enough room: all the **habitats** were occupied by dinosaurs and other big reptiles who ruled at the top of the food chain. With no dinosaurs to compete with, mammals grew much larger, new species appeared and the planet became more diverse. Even if dinosaurs still roamed the Earth, we couldn't coexist: humans would be hunted by the giant meat-eaters among them!

UNCOVERING THE PAST

Can we see live dinosaurs today? Yes, if we count birds as dinosaurs. No, if we're thinking about the big dinosaurs in this book. In this case, we must rely on fossil evidence to tell us what they looked like. Since the 1820s, when the first dinosaur fossils were found, we've been building up our knowledge piece by piece (literally).

Our knowledge of dinosaurs is far from complete: it's very rare for scientists to uncover fossils. Not only are fossils usually deep underground, but it's rare for land-living animals to fossilise in the first place. To fossilise, the animal needs to have been buried quickly, otherwise bones get scattered and rot away from exposure to weather and **bacteria.**

One of the rarest fossils is that of the Oviraptor. A recent discovery of a baby Oviraptor skeleton in its egg has caused a huge leap in our understanding of how these baby dinos hatched.

Fossils can be discovered in different circumstances. Usually, we only find isolated fossil bones which have been separated from the rest of the skeleton. They're not entirely useless - even an isolated tooth or limb bone may be identifiable to a species. However, because the bone has spent millions of years in the ground, it's often weathered or fragmented, making it tricky to find out which dinosaur it came from.

Troodon was one of the first dinos to be discovered in North America. It was found in the 1850s when a palaeontologist found a batch of fossil teeth.

Only very occasionally are articulated skeletons uncovered – this is when the bones are still joined together, as they were in life. More commonly, associated skeletons are found – this is when the bones are jumbled up, but it's obvious that they came from the same animal. It takes a very knowledgeable scientist to put the bones back together.

Group of associated skeletons.

ANYONE CAN FIND A FOSSIL!

You don't need to be a professional palaeontologist to discover a prehistoric creature. Many have been found by children. In 2021, a young girl discovered a 220-million-year-old dinosaur footprint! Given that dinosaurs appeared 230 million years ago, this footprint must be from one of the earliest dinosaurs to walk the Earth.

Perhaps the greatest fossil hunter of all was 12-year-old Mary Anning, who found and **excavated** the first complete skeleton of the prehistoric marine reptile **Ichthyosaurus** in England, in 1811.

Why not begin your own searches by joining a fossil-hunting group?

Statue of Mary Anning, fossil hunter in Lyme Regis, Dorset (England).

INDEX

A
Anning, Mary 31
asteroid 6

B
Baryonyx 9, 10-11, 25
bear 10-11
bird(s) 6, 13, 16-17, 18, 30
bone 10, 15, 16, 29, 30-31

C
cassowary 16
cats 25, 26, 29
carnivore (meat-eater) 7, 8-9, 20, 22, 24-25, 26-27, 28-29
cheetah 7
Coelophysis 8, 20-21, 25
Ceratosaurus 8, 14-15, 23
claw 9, 10-11, 12
Cretaceous period 8-9, 10, 12, 16, 18

E
Edmontosaurus 28
eggs 9, 13, 16-17, 30
extinction 6, 9
eyes 8, 26-27

F
fossil 6-7, 9, 10, 19, 29, 30-31
fox 20-21

H
herbivore (plant-eater) 6-7, 8-9, 12, 23, 24, 26-27, 28
horses 18-19
human impact 24
hunting (hunt) 10, 14, 18, 20, 22-23, 26, 29

I
Ichthyosaurus 31

J
Jurassic period 8, 14, 23

K
kangaroo 12-13

L
lions 14-15

O
Oviraptor 9, 16-17, 30
omnivore 12, 16, 18

P
piscivore 10

R
red crabs 16-17

S
sauropod 14, 22, 25, 28
Sauroposeidon 7
Sir Richard Owen 6

skeleton 10, 16, 30-31
 articulated skeleton 31
 associated skeleton 31
skull 15
Struthiomimus 9, 18-19, 24

T
teeth 9, 10, 12, 14-15, 28, 31
theropod 15
Triassic period 8, 20
Troodon 9, 12-13, 25, 27, 31

V
Velociraptor 27

Z
zebra 22
zoo 25

GLOSSARY

Bacteria - Microscopic organisms that can cause disease.

Carnivore - An animal that eats only meat.

Excavate (verb) - The careful removal of earth from an area in order to find buried remains.

Fossil - The remains or impression of a prehistoric plant or animal embedded in rock and preserved.

Habitat – The place or environment where a plant or animal naturally lives or grows.

Herbivore - An animal which eats only plants.

Ichthyosaur - One of a group of swimming reptiles from the Mesozoic Era. They had streamlined fish-like bodies and tail fins.

Juvenile – An animal which is young - not yet a full-grown adult.

Mass extinction - An event that brings about the extinction of a large number of animals and plants. There have been about five mass extinctions in the history of life on Earth.

Meteorite - A rock from space.

Mesozoic Era – The era of time in which dinosaurs lived, among other animals. It lasted around 186 million years, from 252 to 66 million years ago.

Oases - Green areas in a desert, where there is water and plants grow.

Omnivore - An animal which eats both plants and meat.

Palaeontology - The study of ancient life and fossils. People who study this are called palaeontologists.

Period - A division of geological time that can be defined by the types of animals or plants that existed then. Typically, a period lasts for tens of millions of years.

Piscivore - An animal which eats mostly fish.

Sauropod – A type of extremely large herbivorous dinosaur that had a long neck and tail, trunk-like legs, but small head.

Scavenger - An animal that feeds off food other animals have killed.

Theropod - A type of bipedal carnivorous dinosaur with long jaws, three-toed hind legs, and small front legs with clawed hands.

Tsunami – An extremely long and high fast-moving sea wave caused by an earthquake or other disturbance.

Vertebrae – The bones that make up an animal's spine.